ONE WATER

AN INTEGRATED AND ADAPTIVE APPROACH FOR RURAL WASTEWATER MANAGEMENT IN THE PEOPLE'S REPUBLIC OF CHINA

DECEMBER 2024

ASIAN DEVELOPMENT BANK

CONTENTS

TABLES AND FIGURES

TABLES

FIGURES

ACKNOWLEDGMENTS

The authors of this publication—Mingyuan Fan, principal water resources specialist of the Agriculture, Food, Nature, and Rural Development Sector Office (SG-AFNR), Asian Development Bank (ADB); and Jianpeng Zhou, ADB consultant—would like to express their great appreciation to the management of ADB's East Asia Department (EARD) and SG-AFNR for their valuable guidance, supervision, and support, most especially to Muhammad Ehsan Khan, director general, EARD; Qingfeng Zhang, senior director, SG-AFNR; and Jiangfeng Zhang, director, SG-AFNR.

For his efforts and insightful comments as peer reviewer, the authors are grateful to Jude Kohlhase, principal portfolio management specialist, Sectors Group.

This publication is based on the consultant's final report under ADB's technical assistance project, Rural Vitalization: Rural Wastewater Treatment and Environmental Management Project (TA 9825-PRC), authored by Jianpeng Zhou. Joe Zhao provided guidance to the authors and reviewed the draft report. Joy Quitazol-Gonzalez facilitated the publishing of this knowledge product, including the final review of the draft report. The authors would also like to thank Christian Fischer and Wencan Yu, ADB consultants, for their assistance.

ABBREVIATIONS

ADB	Asian Development Bank
IWRM	integrated water resources management
PRC	People's Republic of China
RWWM	rural wastewater management
SDG	Sustainable Development Goal

ABSTRACT

One Water is an integrated water management approach that recognizes the different types and sources of waters, and the interconnectivity of these resources. It takes a holistic view of a rural community's water resources and assimilates each element into one overall ecosystem. Given that there is no one-size-fits-all solution for rural wastewater management (RWWM), the One Water approach establishes site-specific criteria to develop cost-effective and sustainable RWWM plans and solutions. These criteria take into account the local circumstances (e.g., economic development and climatic conditions), technical feasibility and economic affordability of selected solutions, the availability of water, the characteristics of rural wastewater, and the operational and technical capability and skills of the rural community.

This knowledge product introduces the concepts and best practices that serve as guiding principles of RWWM in the People's Republic of China (PRC). The objectives are to (i) identify the needs and challenges of rural communities in the PRC, (ii) show the diverse characteristics of rural wastewater in the PRC, (iii) introduce the One Water approach for RWWM, (iv) describe One Water's rural wastewater disposal-based discharge criteria, (v) discuss the integrated and adaptive approach for rural wastewater technology selection, and (vi) recommend directions of future work for RWWM.

For this knowledge product, RWWM considers wastewater from both the PRC's administrative villages and natural villages, and focuses on domestic wastewater from residential households. When selecting treatment technologies of domestic wastewater, it is important to choose solutions that are sustainable, maximize the benefits to the rural community, and protect the rural environment.

INTRODUCTION

Background

Rural communities play a crucial role in our society, environment, and economic growth. The well-being of rural communities depends on access to clean and affordable water and sanitation, which are essential for improving the quality of life and protecting the rural water environment. But most households in rural areas still do not have access to basic water and sanitation services.[1] The global cost of providing universally and safely managed sanitation by 2030 in rural areas alone is estimated at $24 billion annually.[2]

> Rural wastewater management is critical for developing the common good of a modern society

Rural wastewater management (RWWM) is critical for developing the common good of a modern society. In fact, RWWM contributes to achieving a range of Sustainable Development Goals (SDGs), such as good health and well-being (SDG 3), clean water and sanitation (SDG 6), decent work and economic growth (SDG 8), reduced inequalities (SDG 10), and sustainable cities and communities (SDG 11).

In the People's Republic of China (PRC), 35% of the country's population, or 500 million people, live in various types and sizes of rural villages.[3] RWWM in the PRC has achieved significant progress in the last 20 years. For example, in 2018, the national government of the PRC initiated the Rural Vitalization Strategy Plan, 2018–2022, which mandates that RWWM for sanitation is to serve as a key component for improving the rural living conditions and reviving the rural ecological environment in the country. The national average rural wastewater service rate increased from 1% in 2006 to 35% in 2020.

Nonetheless, many PRC villages, especially those in underdeveloped and water-scarce regions, still lack access to suitable rural wastewater facilities. Furthermore, many installed rural wastewater facilities do not have adequate technical support and sufficient funding to sustain their operation and maintenance, and/or are unable to meet the technical criteria and standards for environmental protection. Much progress on both the knowledge and practice of RWWM, therefore, needs to be developed and disseminated.

---

[1]   World Health Organization (WHO)/UNICEF Joint Monitoring Programme for Water Supply, Sanitation and Hygiene. https://washdata.org/; and Inter-American Development Bank. 2023. A Call to Action for Better Rural Water, Sanitation and Hygiene Services. Blog. 20 March. https://blogs.iadb.org/agua/en/a-call-to-action-for-better-rural-water-sanitation-and-hygiene-services/.

[2]   United States Agency for International Development (USAID). 2020. Rural Sanitation. *USAID Water and Development Technical Series.* Technical Brief No. 2. https://www.globalwaters.org/sites/default/files/usaid_rural_sanitation_tech_brief_2_508_2_updated.pdf.

[3]   World Bank. Open Data. Rural Population (% of total population) – China. https://data.worldbank.org/indicator/SP.RUR.TOTL.ZS?locations=CN (accessed 4 July 2024).

Objectives and Scope of Work

The purpose of this Asian Development Bank (ADB) knowledge product is to introduce the concepts, international perspectives, best practices, and guiding principles for RWWM. It aims to raise thought-provoking questions, encourage dialogue, and disseminate knowledge, with a focus on the thematic topic of RWWM in the PRC. The specific objectives of this knowledge product are to

(i)   identify the needs and challenges of rural communities in the PRC,

(ii)   show the diverse characteristics of rural wastewater in the PRC,

(iii)   introduce the One Water approach for RWWM,

(iv)   describe One Water's rural wastewater-disposal-based discharge criteria,

(v)   discuss the integrated and adaptive One Water approach for rural wastewater technology selection, and

(vi)   recommend directions of future work for RWWM in the PRC.

ADB recently published two reports on RWWM: (i) *People's Republic of China: Rural Wastewater Management Practice Guideline* (and its related online open access website), prepared under the technical assistance to the People's Republic of China for the Rural Vitalization—Rural Wastewater Treatment and Environmental Management Project;[4] and (ii) *Sustainable Rural Wastewater Management in the People's Republic of China: Institutional, Regulatory, and Financial Frameworks and Stakeholder Participation.*[5] This knowledge product serves as a complementary document to, and should be used in conjunction with, these reports.

This knowledge product also introduces One Water as an integrated and adaptive approach that focuses on RWWM technical solutions and applications. Although it is focused on the PRC, there are guiding principles that may also be applicable to other countries in Asia and the Pacific, and those in other regions.

---

[4]   ADB. 2023a. *People's Republic of China: Rural Wastewater Management Practice Guideline.* PowerPoint presentation of consultant's report (TA 9825-PRC). https://v4.cecdn.yun300.cn/100001_2203015083/RWWM%20Practice%20Guideline%20Introduction_2023.pdf; and Innovative Rural Wastewater Management. China Rural Wastewater Management Practice Guideline. https://www.irwwm.com/.

[5]   ADB. 2023b. *Sustainable Rural Wastewater Management in the People's Republic of China: Institutional, Regulatory, and Financial Frameworks and Stakeholder Participation.* https://www.adb.org/sites/default/files/publication/927661/sustainable-rural-wastewater-management-prc.pdf.

Defining Rural Communities

The PRC's National Bureau of Statistics places communities in two major categories: (i) city and county, and (ii) rural community. Rural communities are defined as communities living in areas outside cities and counties, and they are further divided into two types of villages:

(i)   An **administrative village** is a fifth-level administrative unit, which is under the administrative levels of township, county, city, and province. There are more than 600,000 administrative villages in the PRC. Such a village serves as the country's fundamental organizational unit for the rural population.

(ii)   A **natural village** is not an administrative village nor a formal administrative unit, but usually has defined boundaries.

Regardless of its type, a village is a clustered community with a population of about a few hundred to a few thousand. Rural communities are also defined as villages that have their major economic and social activities in agriculture. Typically, these rural villages are located close to lands for crops, fruits, vegetables, forest, horticulture, and rural roads.

An effective and practical RWWM plan needs to consider the common characteristics of rural communities, such as a small residential population, low population density, and agriculture and related activities, as the main drivers in the economy. The large disparities in factors, such as varying population density, the number of people in a household, privately owned lots versus publicly shared lands, and the administrative structure of the rural community, also need to be addressed.

This knowledge product addresses the RWWM from, and the sanitation needs of, both administrative villages and natural villages.

The Needs of Rural Wastewater Management

Wastewater of rural communities can come from a variety of sources, including residential households; rural institutions (e.g., schools, hospitals); rural economic activities, such as livestock farming; and stormwater runoff that are contaminated by residuals of land-applied fertilizers, pesticides, animal slurry, or irrigation water.

Agriculture activities produce a large amount of liquid and solid wastes. These wastes include semisolid manure, liquid slurry, wastewater, and cleaning or disinfection chemicals from cattle and piggery operations, slaughtering activities, and dairy farming. Other examples are silage liquor from fresh or wilted grass that are made into semifermented product used as winter forage for cattle and sheep, and vegetable washing water that are contaminated by soil and residual pesticides.

Stormwater runoff produces nonpoint source pollution from sediments caused by soil erosion, nutrients, and pesticides. Nitrogen and phosphorus are major pollutants found in runoff, resulting from the application of commercial fertilizer, animal manure, wastewater effluent, and wastewater sludge. Pesticides are widely applied on farmland for pest control to enhance production.

To protect the water environment, rural communities need to consider all types of wastewater. However, for this knowledge product, the focus is on the management of domestic wastewater from rural residential households resulting from blackwater and greywater.[6] The management and treatment of domestic wastewater directly affect the sanitation and hygiene of the people living in the rural communities.

A variety of aspects needs to be considered for RWWM, such as (i) wastewater characteristics—e.g., constitutes and concentrations of pollutants, flow rates, and their variations; (ii) collection and conveyance of the wastewater from individual houses; (iii) wastewater treatment processes and technologies; (iv) residuals from wastewater treatment facilities; and (v) the final disposal of the treated wastewater, and the applicable discharge criteria for the final disposal. RWWM has been a challenge because wastewater flow rates from rural households are typically low, but with large variations due to the low density of the rural population, dispersed siting of houses, and seasonal changes in the number of residents.

Regional Variations in Climatic and Site Conditions

Although domestic wastewater generated from rural residential houses is expected to have similar constitutes (in terms of pollutants, concentrations, flow rates, and their variations), the collection, conveyance, and suitable options for the final disposal of treated wastewater can have large differences among rural communities located across different geographical areas in a large country like the PRC. RWWM needs to incorporate adequate considerations about the impact of such factors and variations on developing practical and cost-effective RWWM plans.

There are different ways in which to categorize the regions of the PRC, such as by the level of economic development, climate condition, land availability, and population density. Each of these factors affects RWWM.

For economic development planning, the PRC provinces are categorized into four groups:

(i)   The eastern region includes Beijing, Fujian, Guangdong, Hainan, Hebei, Jiangsu, Shandong, Shanghai, Tianjin, and Zhejiang.

(ii)   The central region includes Anhui, Henan, Hubei, Hunan, Jiangxi, and Shanxi.

(iii)   The western region includes Chongqing, Gangsu, Guangxi, Guizhou, Neimonggu (Inner Mongolia Autonomous Region), Ningxia, Qinghai, Shannxi, Sichuan, Xinjiang, Xizang (Tibet Autonomous Region), and Yunnan.

(iv)   The northeastern region includes Heilongjiang, Jilin, and Liaoning.

---

6   Blackwater is wastewater from toilets and bathrooms containing urine and fecal matter, as well as water from dishwashers and kitchens that are contaminated by pathogens and grease. Greywater, on the other hand, is wastewater from showers, bathtubs, washing machines, and sinks that contains lower levels of contamination.

The variations in economic development, industries and commerce, and social and educational infrastructure among these regions all affect the financial strength and technical resources of the governments and administrative authorities of the rural communities. These factors influence the planning, engineering design, technology selection, and operation and maintenance of the RWWM. For example, the rural communities in the eastern region, such as those in Jiangsu and Zhejiang provinces, have generally shown stronger economic conditions and have more financial resources for RWWM than other regions—often contributing to better RWWM practices and outcomes.

The climatic condition affects the operation of rural wastewater facilities, the practical options for the final disposal of treated rural wastewater, and the criteria for the receiving water environment. Two main aspects of climatic condition are precipitation and temperature.

> Climate conditions can have large impacts on the planning, engineering, and operation of rural wastewater management

Climatic conditions can have large impacts on the planning, engineering, and operation of RWWM:

(i)   The quantity and characteristics of rural domestic wastewater are affected by the portable water supply that is related to the availability of and access to water sources. In regions of large annual precipitation, rural communities are more likely to have better accessibility to sufficient and affordable portable water supply, which can result in larger per capita wastewater from households. The need for water conservation, and the potential and the adoption of rainwater harvesting, all affect the characteristics of rural wastewater.

(ii)   Treatment requirements of rural wastewater depend on the disposal of the treated wastewater, as the discharge criteria are associated with site conditions for land disposal and receiving water conditions for water disposal. Land disposal of treated wastewater is affected by groundwater tables and the amount of water in soils. Discharging treated wastewater into surface waters depends on the quantities and qualities of river water flows or lakes.

(iii)   The potential of land application of residual solids (i.e., wastewater sludge) from rural wastewater treatment facilities depends on the characteristics and variations of areal precipitation and the frost conditions of soils, which vary across the PRC regions.

(iv)   Temperature variations across the PRC affect the design, technology selection, performance, and effectiveness of the wastewater treatment. Biological treatment is especially susceptible to wastewater temperature, which is affected by the temperature of the ambient airs.

III CHARACTERISTICS OF RURAL WASTEWATER IN THE PEOPLE'S REPUBLIC OF CHINA

The planning and design of RWWM are affected by the quantity and quality of the wastewater. This knowledge product focuses on domestic wastewater—i.e., wastewater coming from urine, feces, and household activities, such as kitchen chores, bathing, and laundry. The impact and uncertainty caused by other sources on rural wastewater (such as wastewater from agriculture, aquaculture, stormwater runoff, and livestock activities) are not covered in this study.

Characteristics Related to Wastewater Quantity

The volume of domestic wastewater produced per person per day is affected by water consumption, which is influenced by many factors, such as potable water supply, climatic conditions (precipitation and temperature), availability of and access to water resources, population size and density of the rural community, level of social and economic development of the community, cost of the supplied water, and the extent of metering and water conservation measures. The characteristics of rural wastewater vary by locality because of the wide differences among PRC regions. Moreover, if a rural community has a sewer system as part of its wastewater management system, stormwater and groundwater can enter the sewers through cracks and leaks as infiltration and inflow, adding to the flow rates of the wastewater.

Domestic water consumption in the PRC's rural communities ranges from 30 liters to 200 liters per person per day (Table 1). Domestic wastewater volume from rural communities is approximately 60%–90% of domestic water consumption, which vary depending on the regions of the PRC (Table 2). If taking 80% of water consumption to estimate the amount of wastewater, this roughly translates to about 24–160 liters of wastewater per capita per day.

The data in Tables 1 and 2 provide compelling information that there is no one-size-fits-all solution for RWWM. The planners, designers, engineers, managers, and citizens in the rural communities need to consider the specifics of a rural community to identify and develop the most appropriate RWWM technical solution. For example, to determine the wastewater quantity of a rural community, the following approach can be used:

> There is no one-size-fits-all solution for rural wastewater management

(i)   Start with historical data of the water consumption and the wastewater flow rate. The data should be representative of the climatic, regional, social, and economic conditions that affect water demand, water sources, and wastewater production.

(ii)   If historical data is not available, data from comparable water and wastewater systems are used. How the comparable systems are selected should be described with adequate details about the applicability of the selected systems.

Table 1: Domestic Water Consumption in the Rural Communities

Type of Villages	Domestic Water Consumption (liters per capita per day)					
	Northeast	Northwest	Southwest	North	Southeast	South–Central
With flush toilets and shower facilities	80–135	75–140	80–160	100–145	120–200	100–180
With flush toilets, but without shower facilities	40–90	50–90	60–120	40–80	90–130	60–120
Without flush toilets and shower facilities	40–70	30–60	40–80	30–50	80–100	50–80

Note: The rural communities covered are those villages with water supply facilities.
Source: Standardization Administration of the People's Republic of China. 2018. *Guidelines for the Treatment of Rural Domestic Sewage (GB/T 37071-2018)*.

Table 2: Ratio of Domestic Wastewater to Domestic Water Consumption in the Rural Communities

Item	Northeast	Northwest	Southwest	North	Southeast	South–Central
Wastewater discharge coefficient (for all wastewater collected)	0.7–0.9	0.7–0.9	0.6–0.9	0.8	0.7–0.9	0.6–0.8

Source: Standardization Administration of the People's Republic of China. 2018. *Guidelines for the Treatment of Rural Domestic Sewage (GB/T 37071-2018)*.

(iii)   If neither historical nor comparable water and wastewater system data are available, a survey should be conducted in the communities of concern to collect relevant data, including a testing program, if possible, and consider using the general available data for the planning and design of the wastewater management solution.

(iv)   Characteristics of the rural community (including population, income, and their growth forecasts) and external factors (such as the impact from major holidays, summer versus winter, and tourism seasons) should also be considered.

Characteristics Related to Wastewater Quality

The key aspects of rural wastewater quality include the composition, mass loadings, and pollutant concentrations of wastewater. The composition of wastewater is affected by variations in climate, social and economic situations, and the habits of the residents. The mass of feces produced by a household is affected by the dietary fiber intake and the number of residents in the household. In regions where water use is low, the concentrations of pollutants in the wastewater are higher than in regions where water use is high because of less water being used for toilet flushing and sanitation.

The RWWM practice guideline for the PRC (footnote 4) describes the typical composition of domestic wastewater from the PRC's rural communities (Table 3). Like the quantity of rural wastewater, rural wastewater quality characteristics in the country depend on the region and the climate of the rural area.

Table 3: Domestic Wastewater Quality of Rural Communities

Parameter	COD	BOD$_5$	NH$_3$-N	TN	TP	SS	pH
Typical values (mg/L)	150–400	100–200	20–40	20–50	2–7	100–200	6.5–8.5

BOD = biochemical oxygen demand, COD = chemical oxygen demand, mg/L = milligrams per liter, NH$_3$-N = ammonia-nitrogen, pH = potential of hydrogen, SS = suspended solids, TN = total nitrogen, TP = total phosphorus.

Source: Government of the People's Republic of China, Ministry of Housing and Urban-Rural Development. 2019. *Technical Standard for Rural Sewage Treatment Engineering (GB/T 51347-2019)*.

As social and economic conditions improve in a rural community, it is expected that the composition and characteristics of rural wastewater would change. Therefore, the planning and design of rural wastewater technical solution need to take into account demographic changes over a design period.

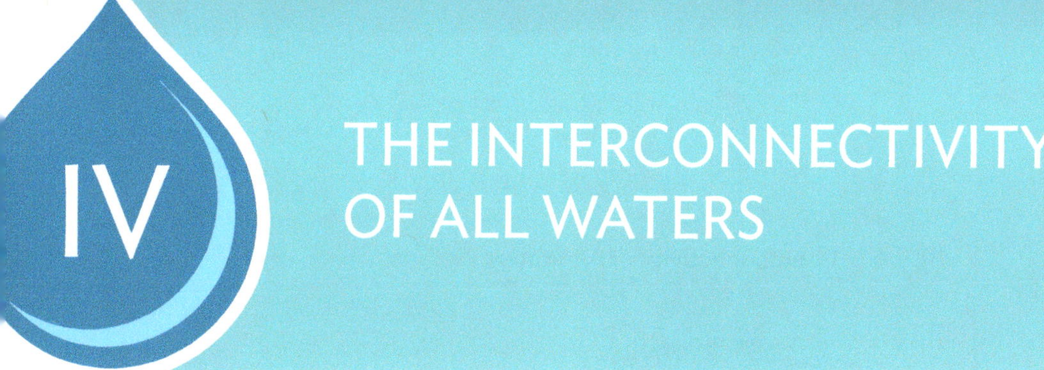

What Is One Water?

Water—in all its forms and notwithstanding its source and end use—has value. This means **all water**, including surface water (seas, rivers, lakes, streams, wetlands, creeks, and reservoirs); groundwater (wells, springs, and aquifers); drinking water; water for household use; industrial and commercial water; water for energy production; rainwater and stormwater; water runoff and floods; and wastewater.

> Water—in all its forms and notwithstanding its source and end use—has value

Traditionally, the water sector has treated and managed drinking water, rainwater, wastewater, and other water sources separately, through its own water systems and processes. One Water seeks to rectify this separation by considering water as a circular, interconnected resource (Figure 1).

By embracing the concept that "all water is one water," the complete life cycle of water can be managed more effectively to maximize water's benefit in building better communities, strong economies, and healthy environments.[7] This is the vision of the One Water approach: all water is a valuable resource and should, therefore, be managed in an inclusive, sustainable, and integrated manner. This holistic approach focuses on sustainability and resiliency, seeking long-term solutions to societal, community, and environmental needs.

Water is life; hence, water is for all. From this belief stems the basic principles of One Water: water equity, water access, and water affordability.[8] **Water equity** means everyone should have equal access to clean, affordable, and safe water. One Water, for example, promotes investments in communities disproportionately affected by water-related challenges to ensure the availability of water resources for all. **Water access** relates to one of the United Nations' sustainability goals for development—i.e., achieving universal access to basic and safely managed water and sanitation by 2030 (SDG 6). According to the United Nations, one in three people worldwide do not have access to safe drinking water, and two out of five people do not have basic handwashing facility with soap and water. **Water affordability** encompasses not just the monetary aspect of water, but also time, health, and social costs, which primarily affect women and children.

---

7   US Water Alliance. 2016. *One Water Roadmap: The Sustainable Management of Life's Most Essential Resource.* https://uswateralliance. org/wp-content/uploads/2023/09/Roadmap-FINAL_0.pdf.

8   C. Tuser. 2021. What is One Water? *Wastewater Digest.* 10 September. https://www.wwdmag.com/utility-management/article/ 10940010/what-is-one-water.

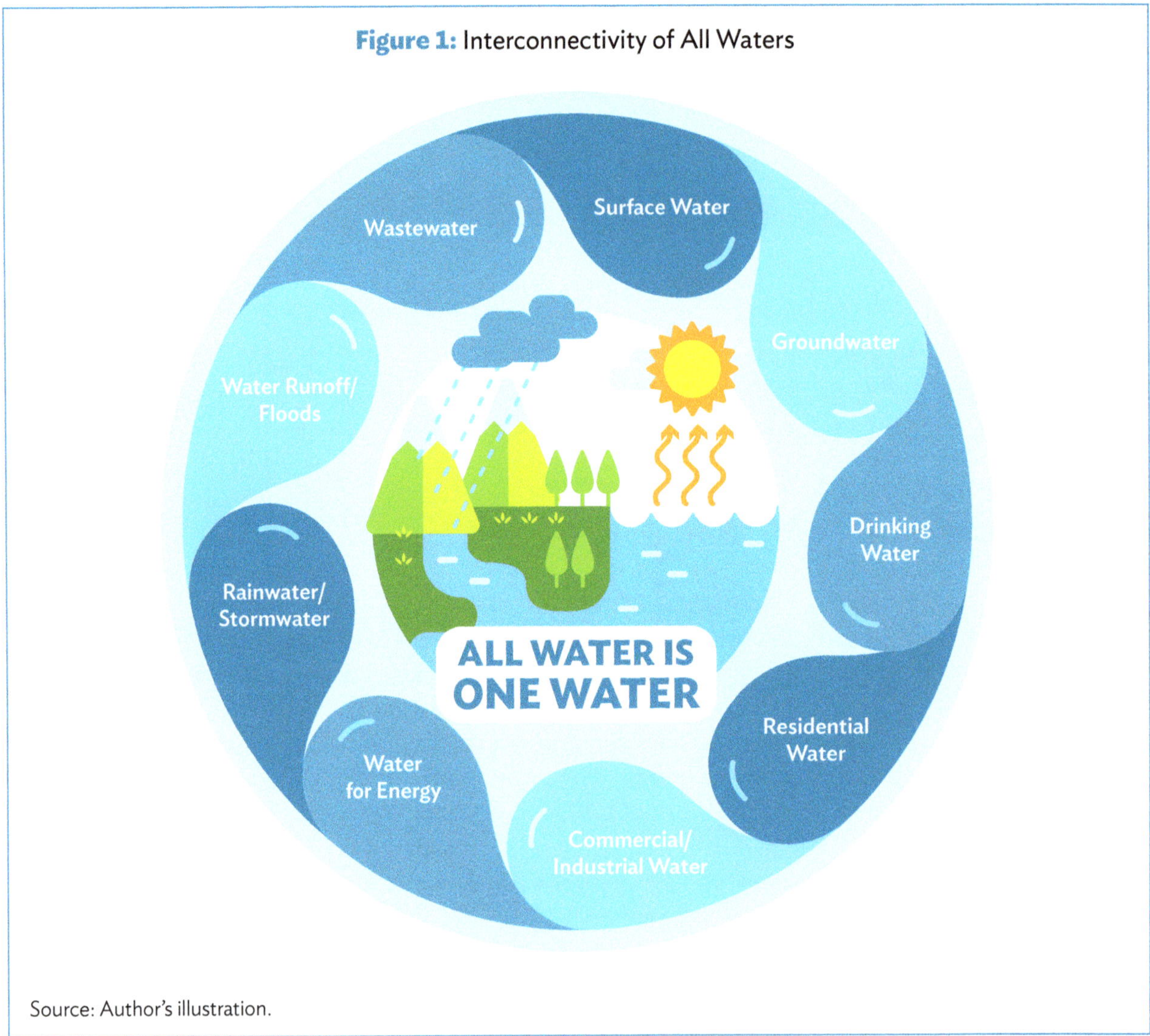

Figure 1: Interconnectivity of All Waters

Source: Author's illustration.

The seven hallmarks of One Water are (i) a mindset that all water has value, (ii) a focus on achieving multiple benefits, (iii) a systems approach, (iv) watershed-scale thinking and action, (v) right-sized solutions, (vi) partnerships for progress, and (vii) inclusion and engagement of all (footnote 7).

One Water is based on the concept of integrated water resources management (IWRM).[9] IWRM presents a means to connect water supply, wastewater, and stormwater with existing water source management. Whereas IWRM is spearheaded by the United Nations as a global initiative, One Water is led by the US Water Alliance as a sustainable water management initiative within the United States.

---

[9]   The Global Water Partnership defines IWRM as the practice of promoting the coordinated management and development of water, land, and related resources to maximize their economic and social benefits in an equitable way without compromising the sustainability of essential ecosystems (International Water Association. Integrated Water Resources Management: Basic Concepts. https://www.iwapublishing.com/news/integrated-water-resources-management-basic-concepts).

One Water and IWRM share the same principles and view the life cycle of water as one water system. They are sometimes used interchangeably to refer to the holistic one water management of water resources. One Water also responds to ADB's Strategy 2030 (i.e., ensuring environmental sustainability)[10] and to the guiding principles of ADB's water sector (i.e., embracing environmental sustainability and circular economy).[11]

Wastewater as a Valuable Water Resource

The water on Earth has an estimated total volume of 1.4×10^{18} cubic meters, which seems fairly abundant in relation to the current average global population of approximately 8 billion people. However, when the waters in the oceans; the waters that are locked in polar ice, soils, rocks, and water vapors; and the waters that are not easily accessible to humans are excluded, approximately 6.6×10^{13} cubic meters (or less than 0.005% of the Earth's total waters) are available for use by human societies. Of this available water, only a small fraction is consumed by humans; most of the water becomes wastewater.

As discussed earlier, all water has value, including rural wastewater (the focus of this knowledge product). Since water can be used, recycled, and reused, turning wastewater into reusable water resources should be explored whenever there are appropriate opportunities.

Under the One Water approach, rural wastewater should not be managed in isolation. Instead, all types of water need to be considered simultaneously to achieve an optimized and sustainable solution that maximizes the benefits to the rural community and protects the environment. In rural communities, domestic water supply; rainwater harvesting and reuse; wastewater collection, treatment, and management; and water conservation are among the key areas that need to be considered for their interconnectivity and related impacts in the development of RWWM plans.

Rural Domestic Water Supply

The supply, consumption, and use of rural domestic water affect the characteristics, flow rate, and variations of rural wastewater, because most of domestic water uses (drinking, cooking, cleaning, washing, bathing, toilet flushing, and others) become wastewater.

Because of the spatial and temporal variations of precipitation, many regions in the PRC experience water shortages and extended droughts, resulting in severe drops of groundwater tables, and drying up of flows in springs, creeks, streams, and rivers. These water shortages negatively affect the domestic water supplies of rural communities, and subsequently affect RWWM. Limited water supply will result in reduced wastewater from households; lower flow rates, if water is used to transport wastes from houses to wastewater treatment facilities; and reduced natural absorption capabilities of the treated rural wastewater.

---

[10]   ADB. 2018. *Strategy 2030: Achieving a Prosperous, Inclusive, Resilient, and Sustainable Asia and the Pacific.* https://www.adb.org/sites/default/files/institutional-document/435391/strategy-2030-main-document.pdf.

[11]   ADB's water sector has five guiding principles: (i) building resilience and adaptive capacity, (ii) promoting inclusiveness and gender equality, (iii) embracing environmental sustainability and circular economy, (iv) improving governance and catalyzing finance, and (v) fostering innovation and technological advancement (ADB. Water. https://www.adb.org/what-we-do/topics/water/overview).

Together, these factors will affect the operation and performance of rural wastewater facilities; thus, they need to be considered in developing RWWM plans.

Rainwater Harvesting and Reuse

Rainwater harvesting means collecting and storing rainwater, instead of allowing it to become stormwater runoff. The collected rainwater from rooftops, impermeable surfaces (e.g., paved areas), or low-permeability areas are stored in tanks, cisterns, or pits for rural sanitation use. Rainwater harvesting offers an independent water source when potable water supply becomes restricted. For rural communities where potable water is costly or difficult to come by, rainwater harvesting and reuse can supplement the water supply, decrease household water costs, reduce demands on well water (which will help sustain groundwater tables), reduce stormwater runoffs, and lessen any runoff-related pollution of freshwater sources. Harvested rainwater may be used for irrigation as well.

Rainwater harvesting can be designed and implemented for different scales, such as at a house, neighborhood, or community level. In regions of the PRC with suitable climate, rainwater harvesting and reuse can provide supplemental waters to meet the rural sanitation demands.

Wastewater Management in Rural Areas

There are several practical options available for RWWM—e.g., (i) collect wastewater from residential houses and treat it to remove pollutants for meeting the discharge criteria; (ii) collect and treat domestic wastewater and reuse it for land application or irrigation; and (iii) build, operate, and manage small low-cost on-site wastewater treatment systems for each house, or a cluster of multiple houses that are near each other.

For the first two options, the RWWM approach needs to include a sewer system to convey the wastewater collected from houses to a wastewater treatment facility to remove pollutants and reduce odors and pathogens. Infiltration and inflow issues of piped networks should be taken into account. These options also need to have a suitable area for disposing the treated wastewater. For the third option, the sites need to have appropriate soil and suitable lot size for an on-site wastewater treatment and disposal system. Trained professionals should operate and manage the collection, treatment, and land application or irrigation systems.

For wastewater management, the conventional sewer system is one of the components that can be costly to build. Servicing a few hundred houses can incur large perpetual annual operating costs. The operation of an open-channel sewer system for rural communities is difficult. Low flows of the wastewater will result in low velocities in open-channel hydraulic conditions and cause unwanted settling of solids from the wastewater in sewers, creating problems of operating the sewer system. This will negatively affect the performance of rural wastewater facilities. Therefore, these options need to be evaluated when developing a cost-effective and sustainable RWWM plan.

Water Conservation

Of the total amount of water consumed in a household, a large fraction of that water becomes wastewater. For RWWM, water conservation for household consumption helps to reduce the produced wastewater and lower the burden on RWWM. Some of the factors affecting water conservation are population density of a rural community, the number and variations of residents in a household, affluence, prices of the potable water, incentives for water conservation, and climates in the region or area of concern. A range of technologies are available for practicing water conservation at the household level, such as (i) low-flush toilets, waterless urinals, or composting toilets; (ii) low-flow shower heads; (iii) dual-flush toilets that use different amounts of water; (iv) faucet aerators, which reduce splashing by using less water yet maintaining wetting effectiveness; (v) rainwater harvesting to lower the use of potable water; and (vi) weather-based irrigation controllers.

For RWWM, active water conservation practices in a rural community will most likely lower the quantity of household wastewater, resulting in increased concentrations of pollutants in the wastewater when pollutant loads remain the same. This affects the planning, design, and operation of rural wastewater facilities, and should be considered when developing an RWWM plan.

One Water Approach for Rural Wastewater Management

One Water is a water management approach that meets the needs of both communities and ecosystems by taking a holistic view of the multiple water resources and integrating each resource into one system. The One Water approach recognizes that different types of water, including drinking water, wastewater, and stormwater, are all valuable resources. The interconnectivity of these resources is considered when dealing with a specific type of water, such as rural wastewater. Furthermore, One Water strives to work with other sectors, such as energy, agriculture, industry, and communities, for achieving inclusive and multibenefit solutions.

The One Water approach (i) considers the linkage among different physical systems across traditional divides; (ii) connects different management and governing agencies across traditional sectors; and (iii) promotes a coordinated solution when addressing problems related to the planning and management of water supplies, wastewater, water quality, stormwater, and watersheds. Managing rural wastewater has transitioned from solving an isolated problem to becoming a cohesive planning and management program for coordinated and aligned actions, changing from a "service-oriented" concept to a "resource management" one. Instead of merely focusing on take–use–treat–dispose of water, initiatives to improve water resource use efficiency has been fully integrated into local economic development. Sustainable sanitation solutions are developed, taking regional economic levels, available water resources, water demand, climatic conditions (including precipitation and temperature), and ecosystem requirements into consideration with circular mindsets, which in turn promote local economic development.

The tools for One Water solutions include water conservation and demand management systems, green infrastructure and stormwater management, fit-for-purpose use of alternative local water sources, recycling of water, management of energy, nutrient recovery, and source separation. These tools can be linked to the circular economy approach, which, like One Water, recognizes water as a precious resource and captures its full value by sustainably managing water resources within the water cycle—i.e., closing the loop on maximizing the use of water resources and reducing waste through reuse and recycling.

It is recognized that One Water can involve many stakeholders, which can make adopting this approach an overwhelming task. Because the goals of One Water contain big-picture and long-term objectives, which often are not compatible with traditional regulation timelines and funding strategies, the implementation of the One Water approach can be a daunting challenge.

Moreover, RWWM needs to incorporate a life-cycle assessment perspective in the planning, technology selection, and long-term operation and maintenance to achieve long-term performance and benefits to the community. Whatever the technologies and processes selected, the impacts on human health (e.g., emissions), ecosystem quality (e.g., pollution reduction), and resources (e.g., energy) should be considered before choosing an optimized solution of RWWM under a given set of conditions. The linkages and consequential impacts of many relevant factors require a holistic approach for sustainable RWWM.

The linkages and consequential impacts of many relevant factors require a holistic approach for sustainable rural wastewater management

The One Water approach varies depending on the needs of the community in question. For this knowledge product, One Water means establishing site-specific criteria to develop cost-effective and sustainable RWWM plans and solutions. These criteria should take into account the local conditions (economic development and climatic conditions), the technical feasibility and economic affordability of selected solutions, the availability of water, the characteristics of rural wastewater, and the operational and technical capability and skills of the rural community.

Final Disposal Sites

Besides providing sanitation and maintaining public health of the residents in the rural communities, a fundamental purpose of RWWM is to protect the receiving water and/or land environment.

Final Disposal Directly to Surface Waters

Treated rural wastewater may be discharged into a surface water body. The treatment requirements depend on the characteristics and assimilative capacity of the water body. The eastern and central regions of the PRC, especially the southeastern region, have an extensive network of surface water, such as rivers and lakes, and have higher annual precipitation for a longer period during the year. Treated rural wastewater is likely discharged to surface waters in these regions.

Surface water can provide dilution of the discharged treated wastewater, which lowers the concentration of pollutants. In addition, surface water typically has self-purification capacities. The concentration of remaining pollutants from the discharged treated wastewater becomes lower because of a variety of processes, such as microbiological metabolism in the receiving water bodies. The magnitude of dilution and the effectiveness of the self-purification process depend on the characteristics of the receiving water bodies. While large and rapid rivers can provide faster and more complete mixing to achieve significant dilution and self-purification, inland lakes may require additional nutrient reduction from the discharged treated wastewater because of eutrophication. Given the variations across regions in the PRC, the flow rates, conditions (e.g., temperatures, pollutant loads), and intended usage of the receiving waters can have large seasonal, year-to-year, regional, and locational differences.

It is crucial, though, to consider the adverse environmental impacts of pollutants in rivers, lakes, and other receiving water bodies as a result of surface water disposal of treated rural wastewater. Wastewater discharge, even those that have undergone treatment, can still introduce a range of pollutants, such as heavy metals, toxic chemicals, and disease-causing pathogens. These pollutants can linger in water bodies, threatening water quality and making it unfit for human consumption, irrigation, and recreational activities.

Final Disposal for Land Surface Application or Irrigation

For a specific rural community, the availability of sufficient and adequate land for the disposal of treated wastewater is critical for effective RWWM. The soil at the land disposal sites should be stable so that it can hold the wastewater discharge. Highly permeable and loose soils allow fast infiltration of the applied wastewater and can cause contamination of the groundwater; thus, they should not be used as final disposal sites. Land surface application of discharged rural wastewater should ensure that the disposal site is at adequate distance from water bodies and are protected. Nonpoint source pollution from stormwater runoff can contaminate surface waters and harm the health and well-being of the local community and the environment. If the groundwater tables are high, the land applied with treated rural wastewater should remain shallow so as not to contaminate the groundwater.

In many rural communities, especially those in the arid or dry regions, treated or partially treated rural wastewater is increasingly used for agricultural irrigation to support growth of crops. The World Health Organization developed guidelines for the safe use of wastewater in agriculture, which advocates a multiple-barrier approach on wastewater use, such as ceasing wastewater irrigation for an adequate number of days prior to harvesting so pathogens could die off in the sunlight and cleaning vegetables with disinfectant before supplying the produce to the market.[12] Because of the varieties of site and climate conditions across the PRC regions, RWWM needs to develop site-specific plans for each application.

Final Disposal for Land Subsurface Discharge

Treated wastewater can be applied to land through subsurface discharge, which uses a subsurface drip system to distribute the wastewater. Such a system typically consists of four mechanisms: a pump tank, a wastewater treatment unit, a filtering device, and a tubing system installed below the ground surface for drip distribution.

Land subsurface discharge can provide nutrients to soil, reduce demand for fresh water for irrigation, recharge aquifers, and keep potential contaminants out of waterways. However, it requires high up-front and ongoing expenses for the purchase and maintenance of land and irrigation equipment. Land subsurface discharge cannot be applied in areas near drinking water sources or with shallow groundwater table to avoid potential pathogen contamination.

The drip distribution system can be used in most site conditions, such as shallow soils, clay soils, and places with moderately saturated conditions. It requires lower land surface area than a surface distribution system, can provide uniform subsurface distribution of the treated wastewater, and can be used on a site of steep slopes. The drip system requires at least 300 millimeters of unsaturated soils below the drip tubing that is typically 150–200 millimeters below the ground surface. The small emitters of the drip system may get clogged with organic matters and solids in the wastewater. The drip distribution method, therefore, requires adequate ongoing service for the operation and maintenance of the drip field.

The land subsurface discharge method is more suitable to arid and dry regions, which often face shortage of water for agricultural irrigation. The discharge criteria depend on the nature of the land use, the proximity to groundwater, and the characteristics of the treated wastewater. These factors can affect rural wastewater planning, implementation, and management.

---

12   WHO. 2006. *WHO Guidelines for the Safe Use of Wastewater, Excreta and Greywater.*

Disposal-Based Discharge Criteria

Many provinces of the PRC have established uniform criteria for treated rural wastewater, such as the discharge criteria to surface waters for urban and rural wastewater treatment in Shanxi Province (Table 4). However, considerations and adjustments of the discharge criteria to suit the local conditions and water quality of the receiving water bodies or environment are not reflected in these criteria.

Table 4: Discharge Criteria of Urban and Rural Wastewater Treatment Facilities in Shanxi Province

Indicator	Discharge Standard of Pollutants for Urban Sewage Treatment Plants				Pollutant Discharge Standard for Rural Wastewater Treatment Facilities in Shanxi		
	1st Level A	1st Level B	2nd Level	3rd Level	1st Level	2nd Level	3rd Level
	Basic Control Indicators				Basic Control Indicators		
pH	6–9				6–9		
COD	50	60	100	120	50	60	80
Suspended Solids	10	20	30	50	20	30	50
Ammonia-Nitrogen[a]	5(8)	8(15)	25(30)	..	5(8)	8(15)	15(20)
	Basic Control Indicators				Optional Control Indicators		
Total Nitrogen	15	20	...	...	20	30	...
Total Phosphorus	0.5	1	3	5	1.5	3	...
Animal and Vegetable Oil	1	3	5	20	3	5	10
Petroleum	1	3	5	15			
Anionic Surfactant	0.5	1	2	5			
Chroma	30	30	40	50			
BOD	10	20	30	60			
Number of Fecal Coliforms	10^3	10^4	10^4	...			

... = data not available, BOD = biochemical oxygen demand, COD = chemical oxygen demand, pH = potential of hydrogen.

[a] The values outside the parentheses are applicable for water temperature >12°C, and the numbers inside the parentheses are for water temperature ≤12°C. Higher ammonia concentration is allowed in the discharged water at lower temperature because of the lower toxicity of ammonia at lower temperature.

Sources: Government of the People's Republic of China, Ministry of Ecology and Environment. 2002. *Discharge Standard of Pollutants for Municipal Wastewater Treatment Plant (GB 18918-2002)*; and Shanxi Provincial Standard of the People's Republic of China. 2019. *Pollutant Discharge Standards for Rural Domestic Wastewater Treatment Facilities in Shanxi Province (DB14/726-2019)*.

As mentioned throughout this knowledge product, there are wide variations among regions of the PRC. This covers the size, demography of rural communities, characteristics of rural wastewater effluent, climatic and site conditions, water use within the community, and wastewater treatment processes. To develop and optimize RWWM, site-specific criteria should be developed. Such an approach affects the planning, engineering, management, and operation of the rural wastewater facilities. This reinforces the need for wastewater discharge criteria to consider local conditions, technical feasibility, economic affordability, and the management and technical capability of operators.

Many technologies have been developed for the treatment of domestic wastewater. Although these technologies can be used for rural wastewater treatment for a specific rural community, the selection of the treatment technologies through an integrated and adaptive One Water approach needs to consider a range of factors, such as site conditions, final disposal options, the handling of residuals, cost–benefit assessment, affordability, and technical and operational capacities of the community. There are no standards or templates that are applicable to all situations for rural wastewater management.

In this knowledge product, the technologies for rural wastewater treatment are grouped into two categories: centralized versus decentralized treatment technologies.

Centralized Rural Wastewater Management System

A centralized rural wastewater management system consists of four major elements: (i) a flush toilet in the house, (ii) a sewer system that collects and conveys the wastewater from multiple households to a treatment facility, (iii) a wastewater treatment facility, and (iv) a final disposal system. The flush toilets, sewer, and final disposal systems used for rural communities are expected to be similar to those used in urban communities (Figure 2).

As illustrated in Figure 3, the technologies for centralized rural wastewater treatment include the following components:

(i)   *Bar screen* captures large items in the wastewater, such as solid objects (sticks, rags, and clothing).

(ii)   *Grit chamber* removes large inorganic materials, such as sand and gravel.

(iii)   *Primary sedimentation tank* removes both floating materials (e.g., oil and grease) and settleable materials (both organic and inorganic).

(iv)   *Secondary biological treatment unit* removes colloidal and dissolved organic matters in wastewater, as well as nutrients, such as nitrogen and phosphorus, as required.[13]

(v)   *Tertiary treatment tank* using, for example, a rotary screen provides an advanced level of purification, filtration, and disinfection to produce water to specification.

(vi)   A *final disposal system* discharges treated wastewater to surface water bodies or to underground soils.[14]

---

[13]   Examples of available biological treatment technologies are activated sludge process (conventional plug flow, complete mixing), trickling filter, sequencing batch reactor, rotating biological contactor, waste stabilization pond, constructed wetlands, and subsurface wastewater infiltration system.

[14]   Metcalf & Eddy, Inc. and AECOM. 2014. *Wastewater Engineering: Treatment and Resource Recovery*. 5th ed. McGraw Hill.

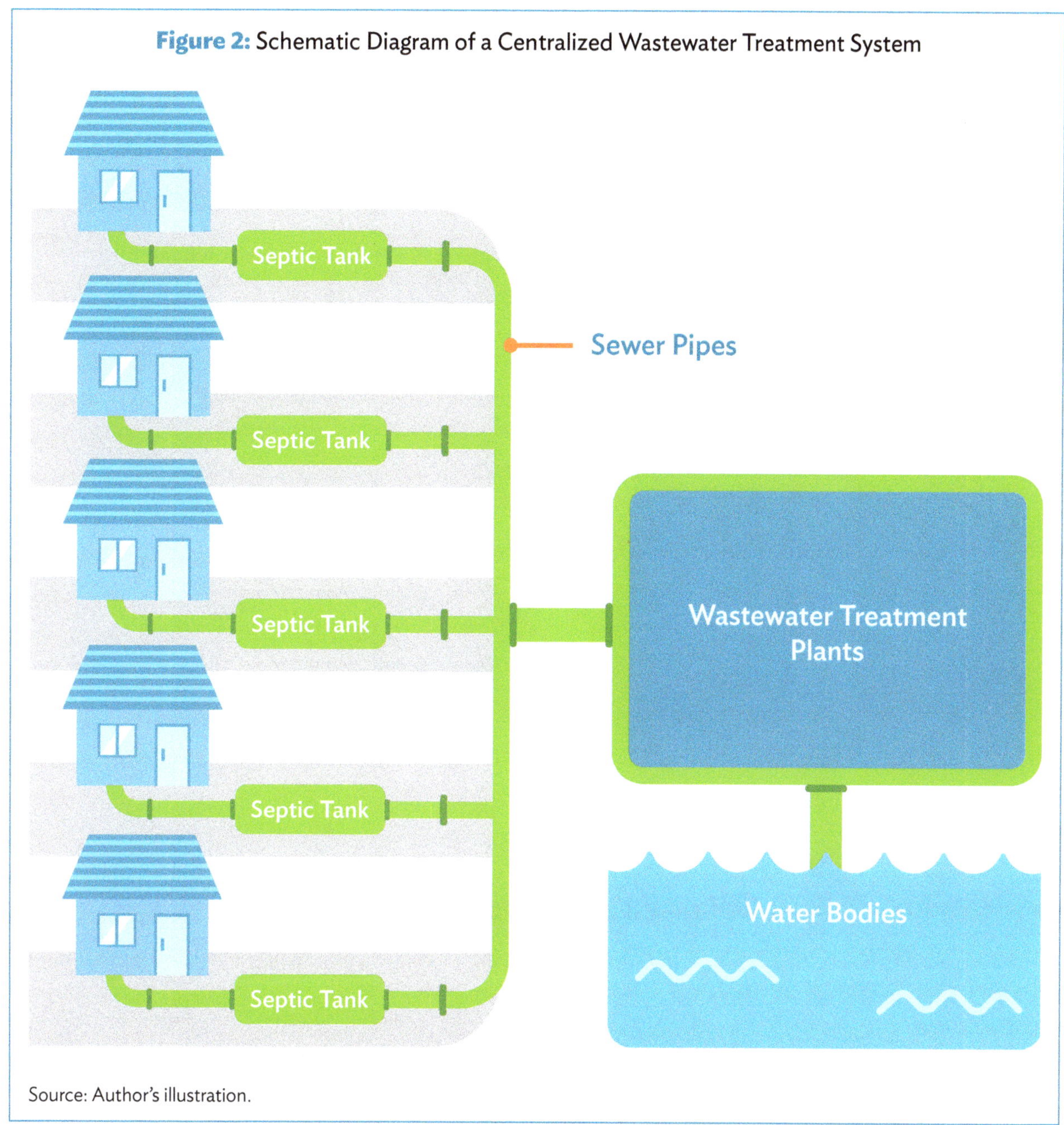

Figure 2: Schematic Diagram of a Centralized Wastewater Treatment System

Source: Author's illustration.

Although many different technologies can be used to treat domestic wastewater, technologies for rural communities should be suitable to handle small flows and large variations in both flows and pollutant concentrations. These technologies should be simple to operate and their operational costs should be relatively low.

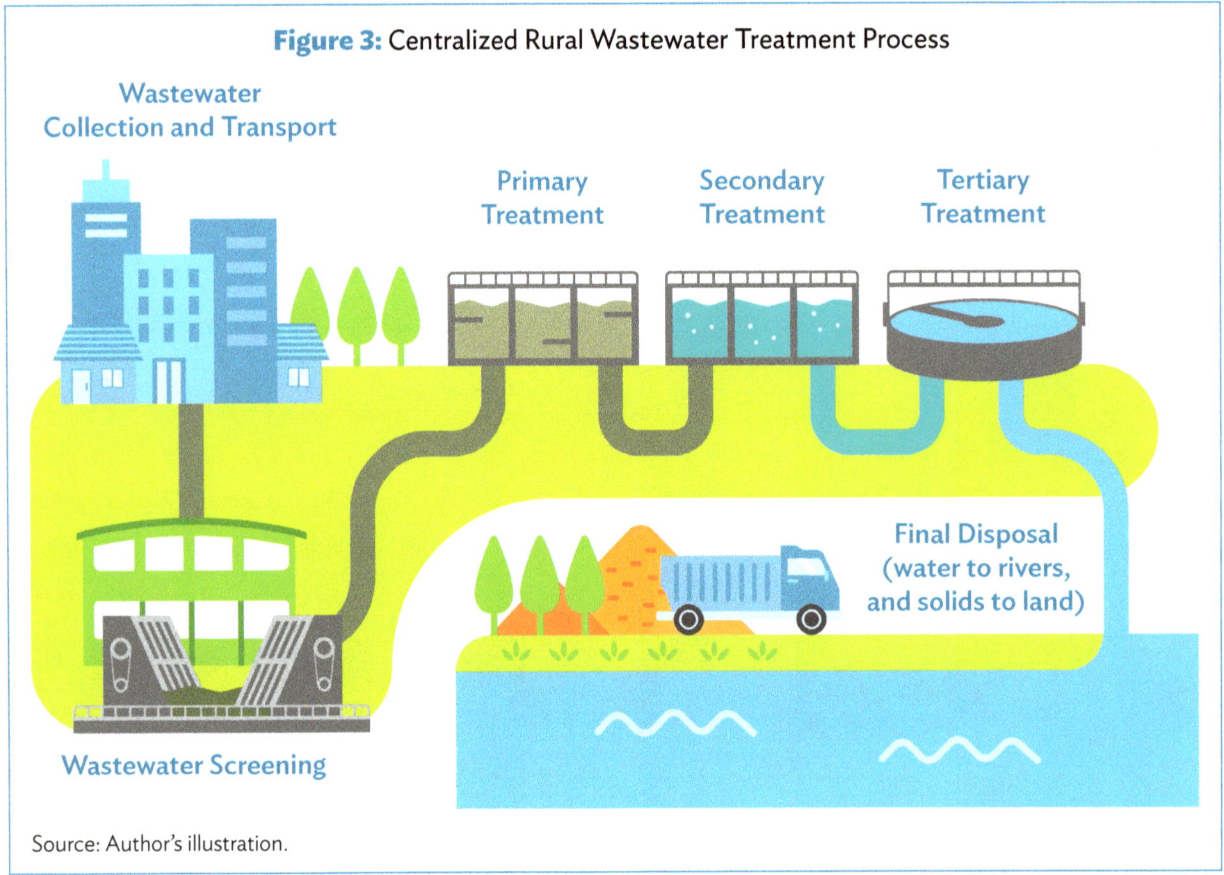

Figure 3: Centralized Rural Wastewater Treatment Process

Source: Author's illustration.

Decentralized Rural Wastewater Treatment System

Technologies for the decentralized treatment of rural domestic wastewater consists of three major elements: (i) a toilet in or near the household, (ii) an on-site wastewater treatment facility, and (iii) a final disposal system.

A range of toilets for rural communities have been developed in the PRC. The PRC's national standard, *Hygienic Specification for Rural Household Latrine (GB/T 19379-2012)*,[15] describes six types of toilets for use in rural communities.

- **Three-chamber septic tank toilet.** This includes a toilet room, a squat pan, a flushing equipment, and a three-chamber septic tank (Figure 4). Land application is the usual method for the final disposal of the treated wastewater from such a toilet. Sediments in the tank are removed as solids. This type of toilet requires water for flushing the feces into the tank.

---

[15]   Government of the PRC, Ministry of Health; and Standardization Administration of the PRC. 2012. *Hygienic Specification for Rural Household Latrine (GB/T 19379-2012)*.

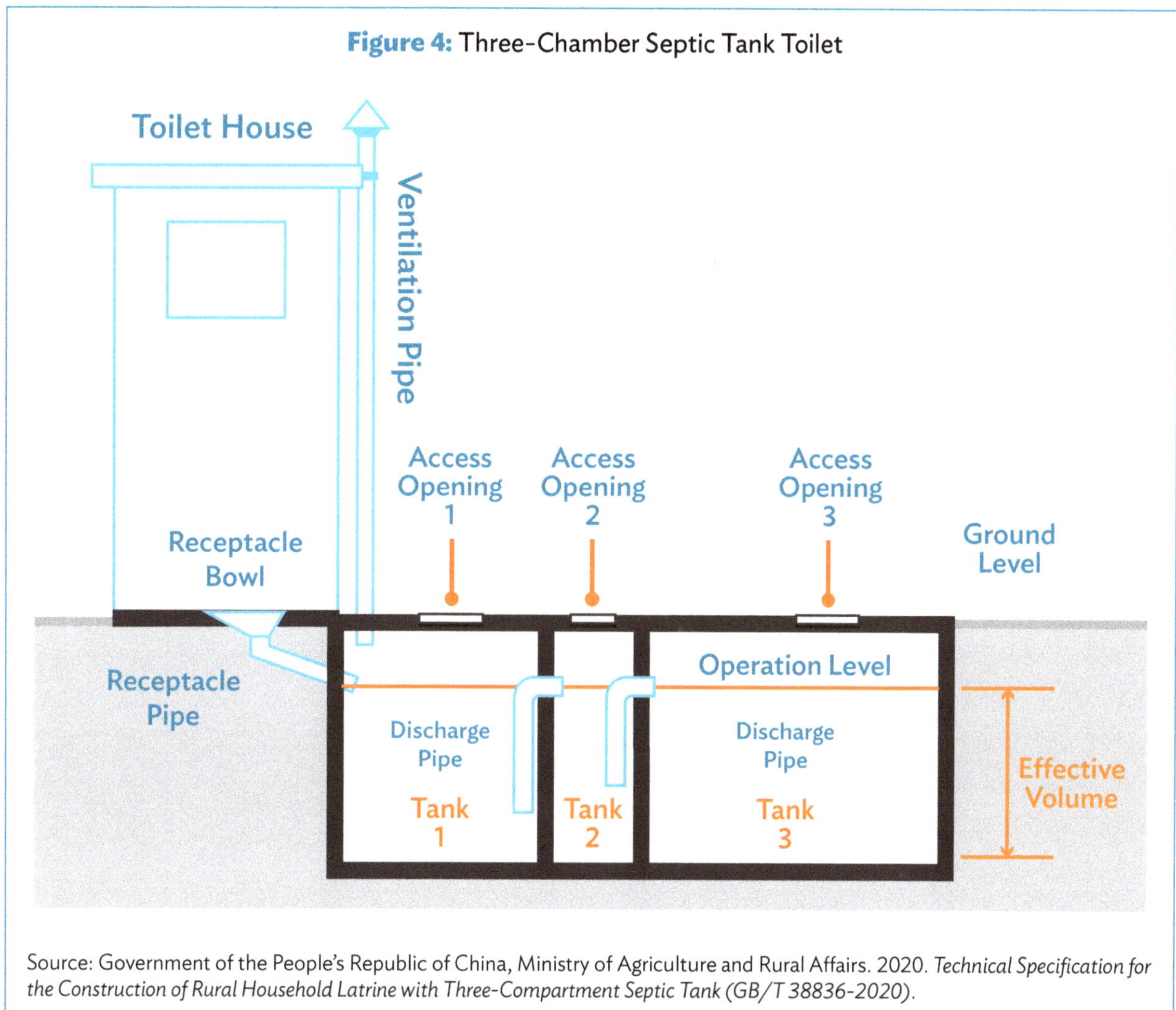

Source: Government of the People's Republic of China, Ministry of Agriculture and Rural Affairs. 2020. *Technical Specification for the Construction of Rural Household Latrine with Three-Compartment Septic Tank (GB/T 38836-2020)*.

- **Dual-urn funneling toilet.** A squatting pan with one outlet is connected to two urns placed underground where feces, urine, and flush water are collected (Figure 5). Although there is no mechanical flush installed, a small amount of water is needed to manually flush wastes.

- **Three-way biogas digester toilet.** This toilet type processes human wastes anaerobically to produce (i) biogas as an energy source for heating and cooking in the house, and (ii) stabilized solid residuals for use as fertilizer in agriculture (Figure 6).

- **Urine-diverting toilet.** It separates the collection of urine and feces (Figure 7). The feces are collected in an underground pit like the double-pit dry toilet, while the urine is collected into a separate urine tank, which can be used as organic fertilizer.

- **Double-pit dry toilet.** It consists of a toilet room and two identical pits for collecting feces (Figure 8). When one of the pits is full, the residents start to use the second pit, while the feces are left in the first pit for decomposition and then removed for land application. The use of the two pits is rotated. The operation of such a toilet does not use any flushing water.

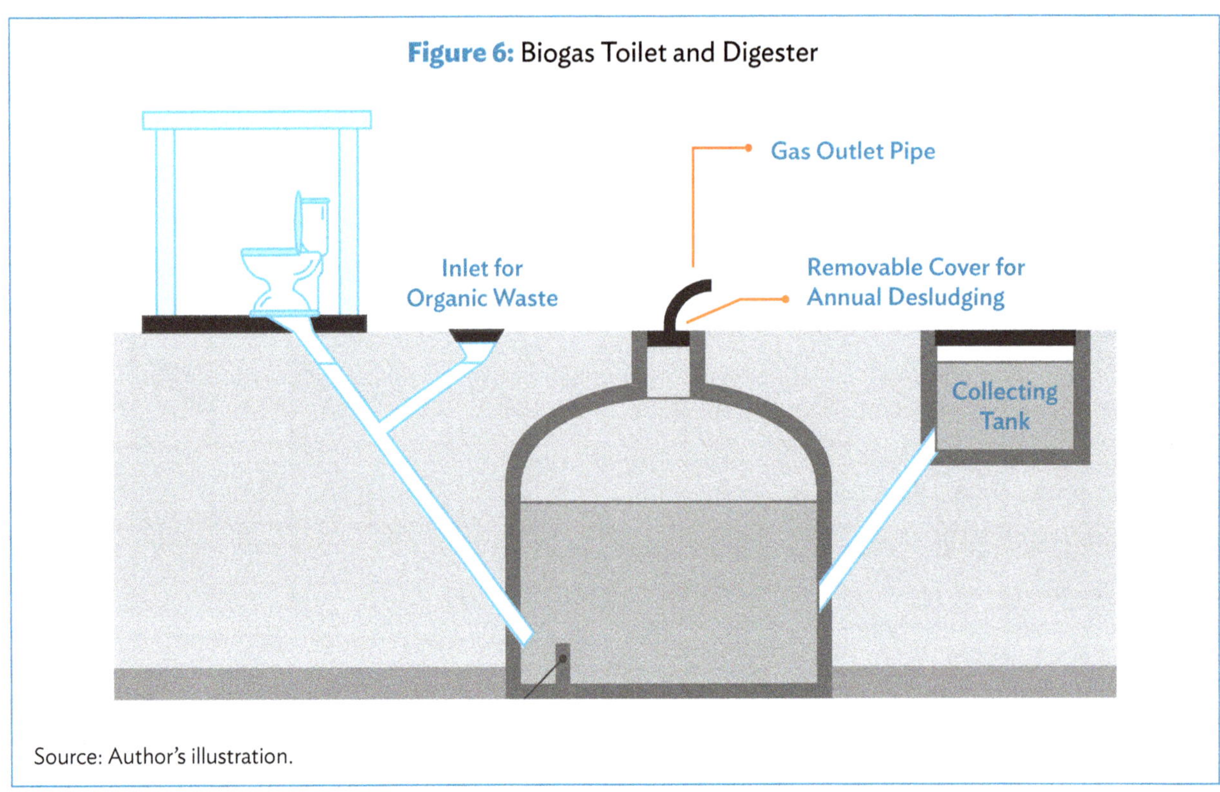

Figure 5: Dual-Urn Funneling Toilet

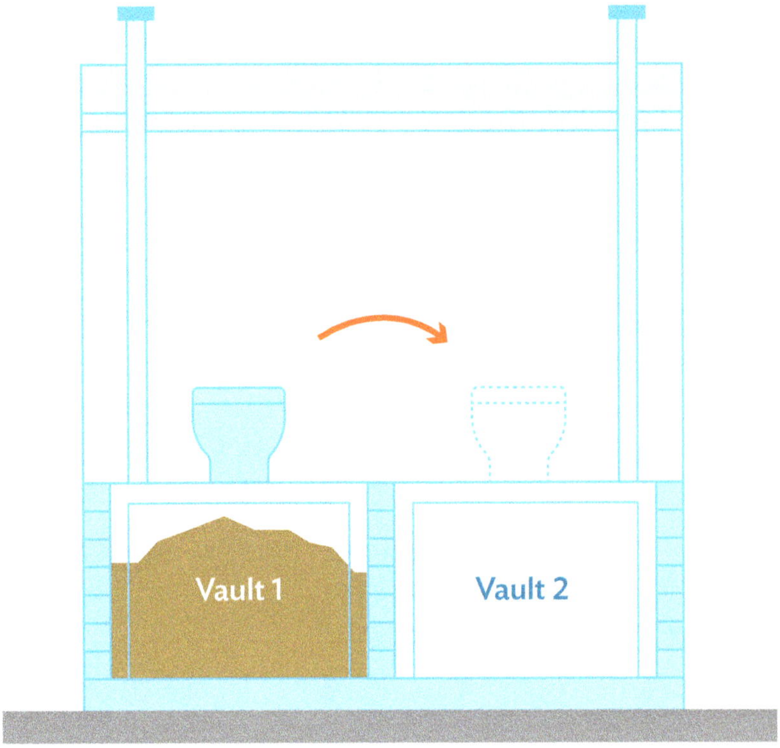

Source: Author's illustration.

Figure 6: Biogas Toilet and Digester

Source: Author's illustration.

Figure 7: Urine-Diverting Toilet

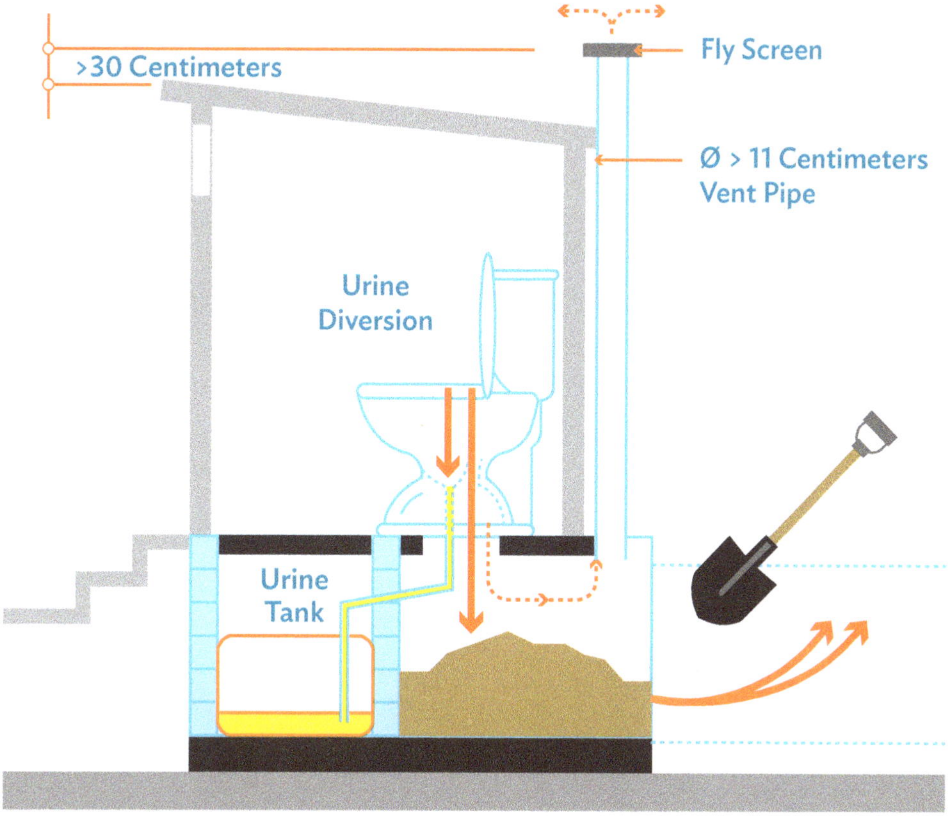

Source: Author's illustration.

Figure 8: Schematic Diagram of a Double-Pit Dry Toilet

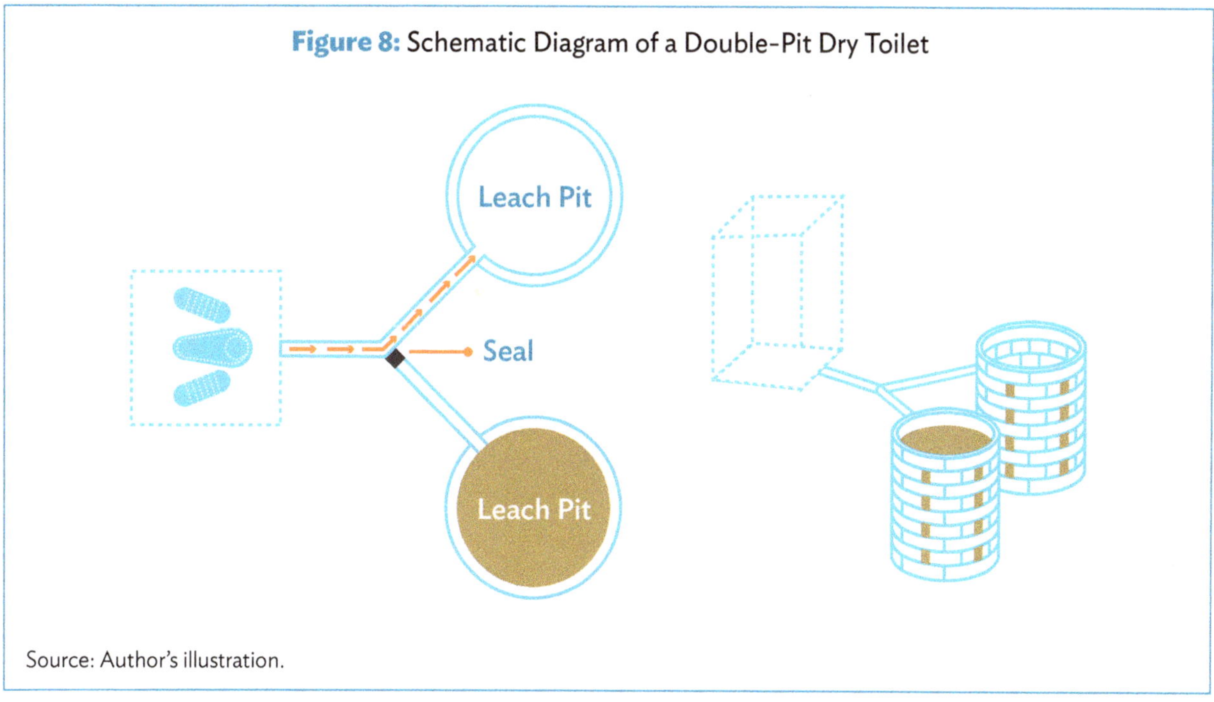

Source: Author's illustration.

- **Flush toilet.** Water is supplied to the toilet for flushing after each use. The produced wastewater is discharged to an on-site septic tank system. A flush toilet may get water directly from the portable water supply system through a cistern in the toilet, or may be operated as a pour flush latrine to which the water is poured into the toilet by the user. Pour-flush latrines use less water than a cistern-flush toilet and require little maintenance.

A septic tank with a field disposal system is commonly used for on-site wastewater treatment and final disposal. Septic tanks are built on-site using concrete block or reinforced concrete, but can also be pre-manufactured with plastic or fiberglass. For a rural community, each house can have its own septic tank, as shown in Figure 9. Alternatively, multiple nearby houses can discharge their wastewater to a large septic tank. All treated wastewater is discharged to land through a single disposal system. Such a cluster system that serves multiple nearby houses is considered as an on-site rural wastewater system.

Figure 9: Septic System with Final Disposal to Land Serving Multiple Houses

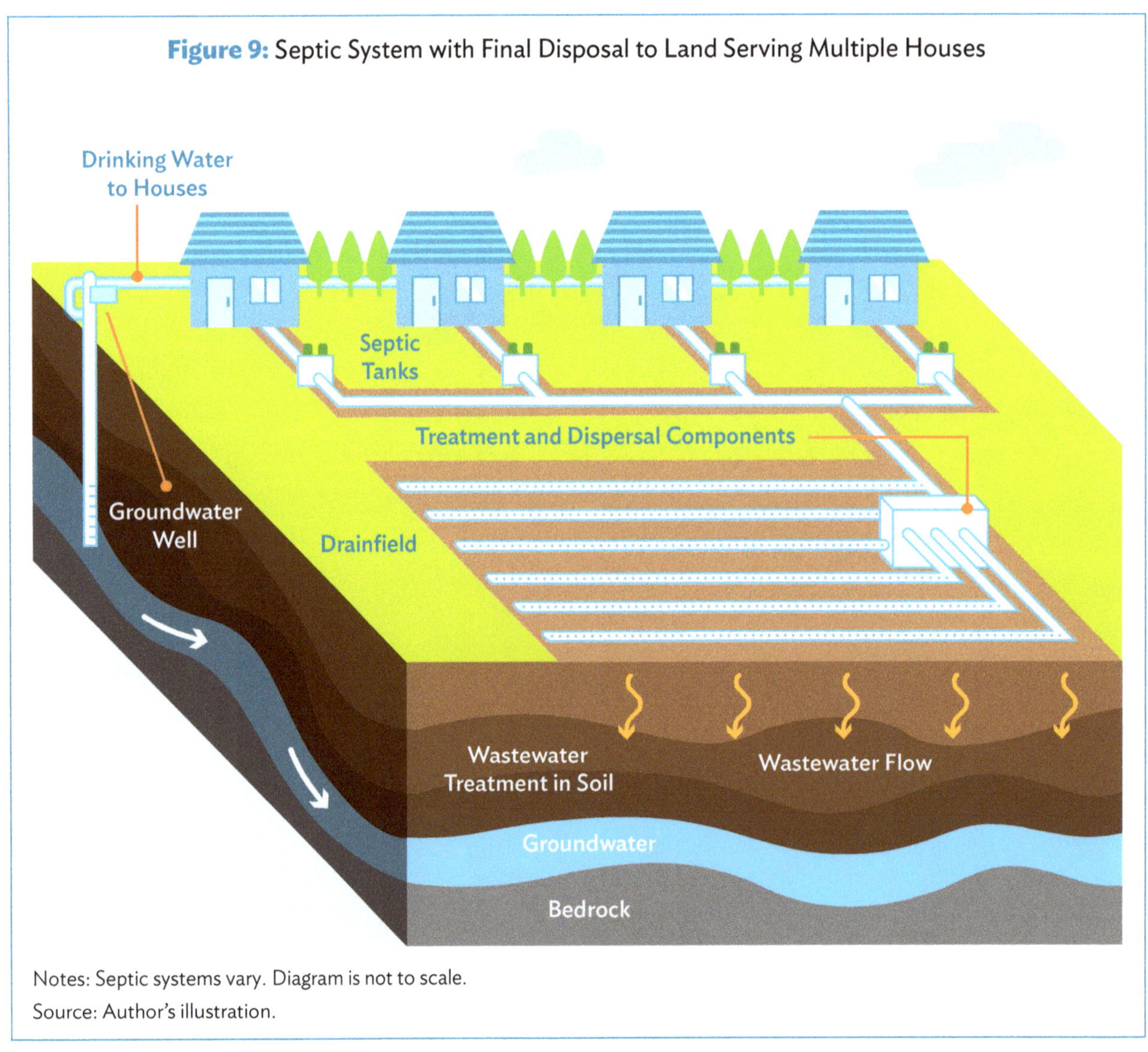

Notes: Septic systems vary. Diagram is not to scale.
Source: Author's illustration.

Supernatants from the septic tank is discharged by gravity or by pumping to a field through a disposal system that consists of a distribution box and perforated piping to a series of shallow and gravel-filled trenches (Figure 10). The field disposal should have suitable soil and groundwater conditions. Adequate separation is required between the bottom of a soil-based disposal system and the groundwater at its highest level. The final disposal field must not cause pollution to groundwater or surface waters. For underdeveloped villages in remote locations or widely spread rural houses, such simple septic tanks with a field disposal system provides a cost-effective RWWM solution.

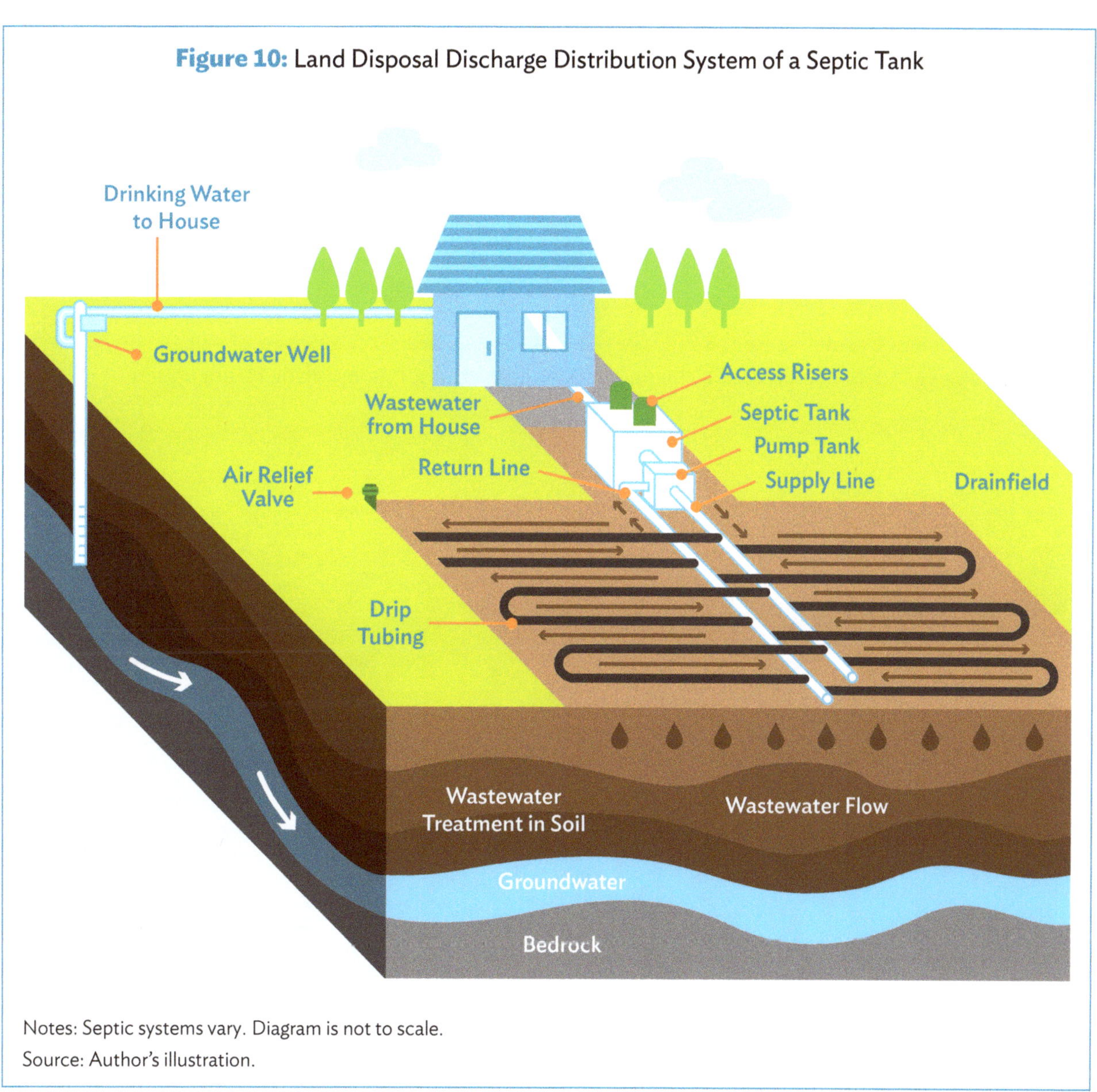

Figure 10: Land Disposal Discharge Distribution System of a Septic Tank

Notes: Septic systems vary. Diagram is not to scale.

Source: Author's illustration.

Disposal of Wastewater Sludge and Biosolids

RWWM, by either centralized or decentralized systems, produces solid residuals as wastewater sludge or as biosolids (i.e., treated sludge). There is a range of treatment and final disposal methods for the handling of wastewater sludge, the most common of which are land application, landfill, and incineration.

Wastewater biosolids may be applied to farmland for agricultural use or treated to produce Class A biosolids that are permitted to be publicly distributed for landscaping, horticulture, agriculture, land restoration, and silviculture. Biosolids produced from wastewater sludge are rich in organic and inorganic nutrients, nitrogen, and phosphorus, which are beneficial for plant growth. Biosolids can be used as fertilizer substitute, as long as their heavy metals content comply with regulated standards.

Biosolids-amended land can improve the physical, biological, and chemical composition and properties of soils; enhance crop yield; and help control erosion. When biosolids are utilized as fertilizer substitute, the resources and energy linked with the raw material extraction (e.g., mining of the phosphorus), manufacturing, packaging, transport, and application of chemical fertilizers are reduced. Therefore, from a life-cycle perspective, biosolids can reduce the environmental impact of fertilizer use and promote best practice of sustainable waste management.

The RWWM practice guideline for the PRC provides basic discussions on wastewater sludge and biosolids, which can be expanded for increased details and added technical solutions in any future update.

VII — DIRECTIONS OF FUTURE WORK

Several external and internal factors need to be considered in developing suitable solutions for managing rural wastewater. The current situation should be evaluated against the anticipated needs for the future. The following are some recommendations of future work for RWWM in the PRC:

- **Develop environmental policies and legislations.** Policies and regulations serve as the basis for the planning and engineering of RWWM. Requirements and standards determine what wastewater management technologies and processes should be selected to meet the need.

- **Promote recovery and reuse of by-products.** Recovering and reusing by-products, like biosolids, as substitute of fertilizers benefit the sustainable management of wastes from rural communities.

- **Evaluate the potential risks of final residuals.** Disposed rural wastewater and sludge can contain residual pollutants, such as toxic materials, pathogens, and decaying organic matters. Risk assessment is needed for the safe disposal of treated wastewater and sludge.

- **Strengthen public acceptance.** The perception of the public about the safety of the final disposal affects the acceptance of a waste management solution. Public acceptance is crucial for sustaining the RWWM practice.

- **Assess the technical feasibility and reliability of the treatment technologies.** From the range of technologies available for rural wastewater treatment, the selected technology needs to be technically feasible, should be able to deliver reliable performance for full-scale application, and can sustain effective and efficient long-term performance.

- **Integrate local economic conditions of the community and affordability of solutions.** The costs of different wastewater treatment solutions can significantly vary across different rural communities, as influenced by the trade-offs in economies of scale and performance. To sustain the long-term financial viability of RWWM solutions, the economic conditions of the rural community and the affordability of solutions need to be analyzed. When developing a sustainable RWWM plan, it is important to conduct a life-cycle cost–benefit analysis to support decision-making.

- **Enhance the skill sets of the management and operational staff.** More often than not, a small rural wastewater treatment facility does not necessarily have access to the same skill sets of staff as a large treatment facility. The complexity and skill set demands of selected waste management technologies should, therefore, be compatible with what is available to the rural community.

- **Design implementation route and phases.** The development and implementation of a waste management plan require substantial capital expenditures, which are usually supported by the capital improvement financing plan of the community. The waste management plan should be compatible with this financing plan and should be designed as a phased program.

REFERENCES

Asian Development Bank (ADB). 2023a. *People's Republic of China: Rural Wastewater Management Practice Guideline*. Consultant's report (TA 9825-PRC).

ADB. 2023b. *Sustainable Rural Wastewater Management in the People's Republic of China: Institutional, Regulatory, and Financial Frameworks and Stakeholder Participation*. https://www.adb.org/sites/default/files/publication/927661/sustainable-rural-wastewater-management-prc.pdf.

Centers for Disease Control and Prevention. Global Water, Sanitation, and Hygiene (WASH). https://www.cdc.gov/healthywater/global/wash_statistics.html.

Innovative Rural Wastewater Management. China Rural Wastewater Management Practice Guideline. https://www.irwwm.com/.

Inter-American Development Bank. 2023. A Call to Action for Better Rural Water, Sanitation and Hygiene Services. Blog. 20 March. https://blogs.iadb.org/agua/en/a-call-to-action-for-better-rural-water-sanitation-and-hygiene-services/.

Tuser, C. 2021. What is One Water? *Wastewater Digest*. 10 September. https://www.wwdmag.com/utility-management/article/10940010/what-is-one-water.